James Alfred Wanklyn, W. J. Cooper

Air-Analysis

A Practical Treatise on the Examination of Air - With an Appendix on Illuminating Gas

James Alfred Wanklyn, W. J. Cooper

Air-Analysis
A Practical Treatise on the Examination of Air - With an Appendix on Illuminating Gas

ISBN/EAN: 9783337139292

Printed in Europe, USA, Canada, Australia, Japan

Cover: Foto ©berggeist007 / pixelio.de

More available books at **www.hansebooks.com**

AIR-ANALYSIS:

A PRACTICAL TREATISE

ON

THE EXAMINATION OF AIR.

WITH

AN APPENDIX ON ILLUMINATING GAS.

BY

J. ALFRED WANKLYN

AND

W. J. COOPER.

LONDON:
KEGAN PAUL, TRENCH, TRÜBNER, & CO., LTD.
1890.

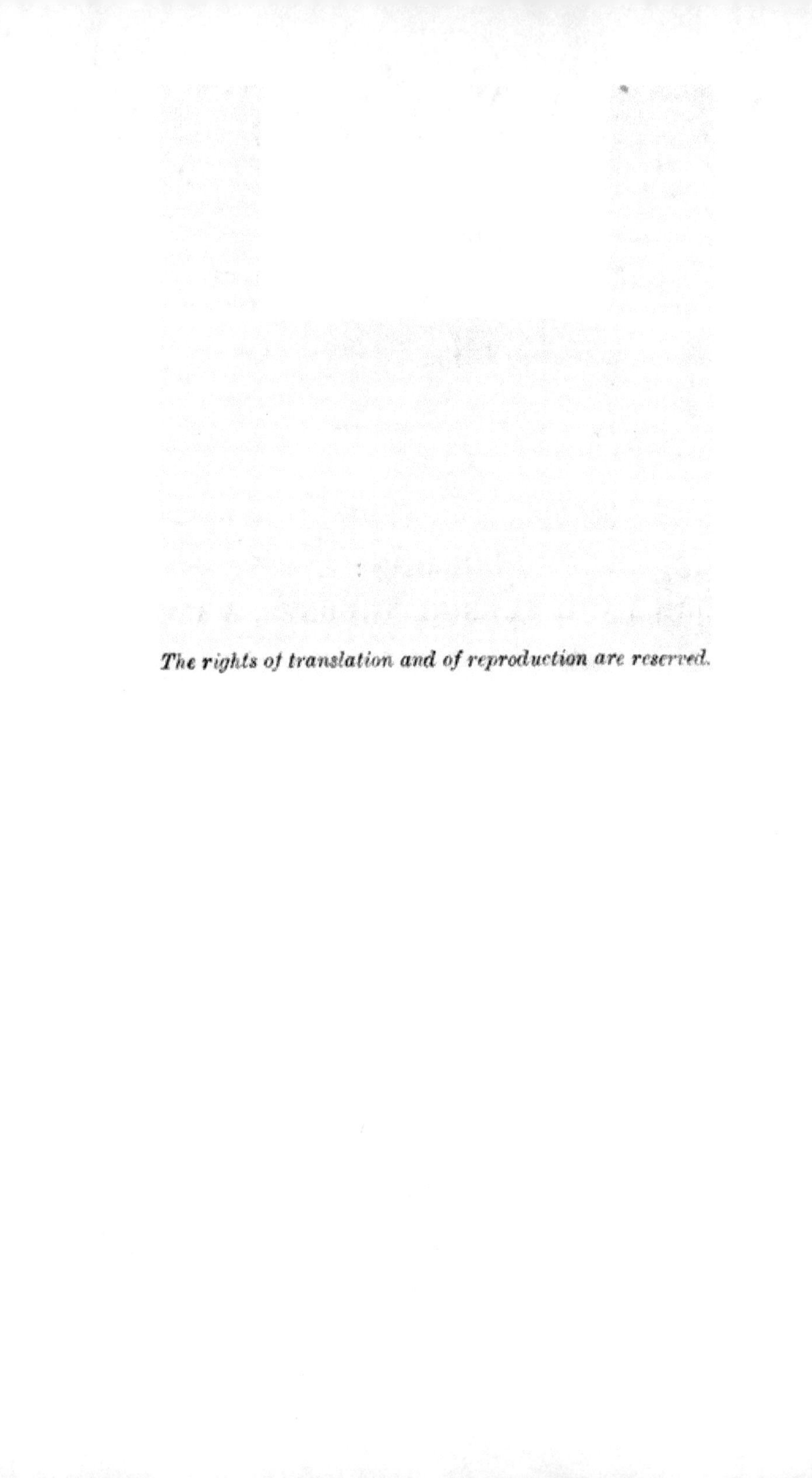

PREFACE.

THIS book is written for the chemist who follows his profession, for the student as a companion in the laboratory, and for the sanitarian.

The charming ease and simplicity and rapidity of much of the work described in its pages will commend it to the chemist, who has hitherto been inclined rather to shun gas-analysis.

The accuracy of modern technical processes of gas-analysis has, we are aware, often been called in question. In particular, it has often been said that measurements over water cannot be accurate because of the solubility of all gases in water. If, in the Hempel apparatus, the large bulk of liquid were boiled out, so as to be devoid of air, there might be reason for the objection. But it is not; and, therefore, the error is restricted to that which arises from the difference of solubility of different gases. It is further to be noted that this error is eluded to a great extent, inasmuch as the contact between the gas and liquids in Hempel's apparatus is too superficial to bring out these differences of solubility. In order to introduce serious solubility error, shaking up might be resorted to, and in that manner possibly some error might arise. But actual handling of the apparatus can hardly fail to bring home to the chemist that it is fairly accurate.

The precision with which 100 c. c. of air measured out in the gas-burette will return from the potash absorption-pipette without failure of 0.1 c. c., or the precision with which a similar experiment may be carried out (as we have often observed) with the well-purified coal-gas of a great London Company, must appeal to a chemist.

The rehabilitation of the Cavendish method of measuring the oxygen in the air will, we think, interest chemists.

The method proposed for the measurement of traces of carbonic oxide in air is original. Other matters which are new will doubtless be found by our readers.

LABORATORY, NEW MALDEN, SURREY,
May 1890.

CONTENTS.

INTRODUCTION.

CHAPTER I.

CHAPTER II.

CHAPTER III.

CHAPTER IV.

CHAPTER V.

APPENDIX.

LIST OF PLATES.

AIR-ANALYSIS.

INTRODUCTION.

(1.) *Nature and General Properties of Air and Gases.*

The eye sees no difference between a bottle full of air and a bottle from which the air has been abstracted. The receiver of the air-pump looks just the same when the exhaustion of the air has been accomplished as it did before the commencement of the pumping. Air, and indeed * *almost* every gaseous body, is invisible.

But, notwithstanding that it is hidden from the eyesight, the air is as real as any other kind of matter with which we come in contact, and manifests itself continually to our senses in the daily course of our lives. Chiefly through its property of *resistance*—an invisible something which resists—does the air make known its presence.

The thorough materiality of that which fills a distended air-cushion or a blown-up bladder is palpable enough to the senses of every person. The reality of the wind appeals to all. It is the pressure of the

* A few gases are coloured, *e.g.*, chlorine and peroxide of nitrogen.

atmosphere which enables the common pump to perform its task of raising water from the well. The pressure of the atmosphere maintains the column of mercury in the tube of the barometer, and the varying pressure of the atmosphere registers itself in the varying height of column of the barometer.

The characteristic property of gases which distinguishes them from liquids and solids is their almost absolutely perfect elasticity. They are not only mobile or fluid (like liquids), but they are elastic, whilst liquids are almost devoid of elasticity altogether. The distinction is far-reaching and fundamental. A cubic foot of water is a fairly definite quantity of water, weighing $62\frac{1}{2}$ lbs. at ordinary temperatures and 2 or 3 lbs. less at the boiling-point of water, and only very minutely influenced by the pressure to which the water is exposed. But a cubic foot of air is a perfectly indefinite quantity, unless the pressure be either expressed or understood.

Under average atmospheric pressure, viz., 30 inches of mercury, a cubic foot of air weighs about 1.3 ounces. Under a pressure of 3 inches of mercury, a cubic foot of air weighs about 0.13 ounce. Under a pressure of 300 inches of mercury, a cubic foot of air weighs about 13 ounces.

In short, the weight of a given volume of air or any gas is exactly proportional to the pressure.

Another mode of expressing the same fact is, that the volume occupied by a gas is inversely proportional to the pressure; and this is the celebrated law of Boyle or of Mariotte.

This law sets forth that reduction of pressure causes expansion of gas, and that increase of pressure causes contraction—a cubic inch of gas becoming two cubic inches when the pressure is halved, and half a cubic inch when the pressure is doubled.

The precise meaning and working of Boyle's law may be learnt by experimenting with a simple piece of apparatus consisting of a glass U-tube, closed at one end and open at the other, and containing mercury, which serves to confine the position of air operated upon, and also enables the pressure to be varied and measured. Experimenting with this apparatus, it will be observed that, in order to compress a volume of air into half its space, it is necessary to supplement the atmospheric pressure by 30 inches of mercury in the open end of the U-tube. On the other hand, diminished pressure is attained by counterbalancing atmospheric pressure by means of a column of mercury in the closed side of the U-tube. If the U-tube be graduated accurately, and if readings of the barometer be made, the mathematical accuracy of Boyle's law may be demonstrated. But without going so far, and without any very great expenditure of trouble, the mind of the operator may be familiarised with that which is meant by the elasticity of gaseous bodies.

Air (and gases generally) expands uniformly under

the influence of heat, and the expansion is greater than the expansion of liquids or solids. In passing from zero centigrade to 100° centigrade, one volume of air expands 0.366. Inasmuch as the expansion is perfectly regular, the relative volumes of the same quantity of air at different temperatures are as follows:—

Degree centigrade.	Volume.
0°	1.00000
1°	1.00366
2°	1.00732
3°	1.01098
4°	1.01464
5°	1.01830
10°	1.03660
100°	1.36600

When air is cooled below zero centigrade, it contracts with regularity at the same rate, the relative volumes of the same quantity of air being:—

Degree centigrade.	Volume.
0°	1.00000
1°	0.99634
2°	0.99268
5°	0.98170
10°	0.96340
100°	0.63400
200°	0.26800
273°	0.00000

The last term is, of course, hypothetical. Before the absolute zero is approached, air and every other gas ceases to exist in the gaseous form, and liquefies and freezes.

If, as has been proposed by some physicists, we were to shift our zero from the freezing-point of water down to *absolute* zero, we should be able to enunciate the law of expansion of gases that the volume of the gas is exactly proportional to the temperature of the gas.

With the existing centigrade scale, however, the law stands thus:—

The figure 0.00366 is the co-efficient of expansion for one degree centigrade, and by that fraction of its volume at zero centigrade the gas expands or contracts for every degree of departure from zero centigrade.

If we desire to compare the quantities of two specimens of air or gas, it is open to us to weigh them and compare their weights. This procedure is sometimes resorted to with advantage; but for obvious reasons, which need not be mentioned here, the weighing of gases is by no means a convenient operation, and the usual method is to measure the gases.

For this purpose three data are required, viz.:—

Observed volume	=	v
Observed pressure	=	p
Observed temperature	=	t

These three data being known, we are in a position to calculate V, which is the volume of the gas at standard pressure and zero—

$$V = \frac{vp}{(1 + .00366\,t)\,P}$$

P is standard pressure, usually put down as 760 millimetres of mercury, but it may be taken as unity, which simplifies the formula—

$$V = \frac{vp}{1 + .00366\,t}$$

In cases where the gases to be dealt with are at the same temperature and under the same pressure, no correction for temperature and pressure is required, and the observed volumes may be directly compared.

In these instances—now rising into prominence and importance—when rapid, and indeed almost instantaneous, methods of analysis are resorted to, the readings of gaseous volumes are made at the temperature and pressure which happens to prevail at the time and place where the analysis is performed. No reductions of such volumes are called for; no record of the thermometer and barometer is required. A convenient phrase describing the procedure is, "Readings at current temperature and pressure."

The gaseous condition of matter is of great scientific interest. Gases, as we meet with them under ordinary conditions at the surface of the earth, are far lighter than liquids or solids. A litre of the lightest known solid, the rare metal lithium, weighs 578 grammes. A litre of the lightest known liquid, viz., butyl-hydride, C_4H_{10}, weighs more, viz., 600 grammes. But a litre of air under the ordinary conditions of pressure and temperature weighs only $1\frac{1}{3}$ gramme; a litre of the heaviest known gases comes up to only about 6 or 7 grammes in weight. Manifestly the gaseous state of matter, as we meet with it, is a state of extreme expansion.

A litre of air consists mainly of intermolecular space, more than 99.9 per cent. intermolecular space, and less than 0.1 per cent. of space totally filled with molecules.

Physicists regard the constitution of gases as being much simpler and more regular than the constitution of liquids and solids. Equal volumes of gases under the same pressure and temperature contain the same number of molecules; from which follows the important consequence that the specific gravities of gases correspond exactly to the molecular weights of the gases.

(2.) *Chemical Composition of the Atmosphere.*

The atmosphere consists mainly of a mixture of the two elementary gases, oxygen and nitrogen, in the following proportion:—

	Volume.
Oxygen	21.0
Nitrogen	79.0
	100.0

It is remarkable that so fundamental a fact as the composition of the atmosphere should have remained unknown until comparatively modern times. At the close of the year 1774 the discovery was made that there is an active constituent in the air, and to this active constituent the name oxygen was given; and the important part played by oxygen in the combustion of burning substances and in the respiration of animals was duly recognised.

At first, and for some years, it was thought that there was great variation in the proportion of oxygen, and peculiarities of climate were supposed to depend upon this variation. Gradually, however, this notion was dispelled, and it has been established beyond dispute that the proportion of oxygen in the air is almost absolutely the same all the world over. Dry air consists of:—

	Volume.	
Oxygen	20.8 to	21.0
Nitrogen	79.2	79.0
	100.0	100.0

Under rare conditions—in India, for instance, where there is much decaying organic matter—it is recorded that the oxygen has fallen to 20.4 per cent.; but that is looked upon as an exceedingly rare occurrence.

Next in abundance to oxygen comes the aqueous vapour—the moisture—of the atmosphere. This is the very variable constituents. During frost the aqueous vapour falls to 0.5 per cent. of the volume of the air, whilst in hot weather it sometimes rises above 3 per cent. of the volume of the air.

Nitrogen, oxygen, and water—these three substances are the major constituents of the atmosphere. Everything else in the atmosphere is present in very minute proportion.

It is, however, with substances present only in small proportion—with minor constituents—that the sanitarian has generally to deal in his examinations of the air we breathe.

The carbonic acid in the air ranges from 0.03 to 0.06 volumes per 100 volumes of air, and is by far the most abundant of the minor constituents.

In relation to the air the carbonic acid may be compared with the mineral matter in ordinary drinking-water. The proportion of carbonic acid in air is about the same as the proportion of mineral matter in drinking-water. The carbonic acid arises from the combustion of fuel, from the decay of organic bodies, and from the ordinary respiration of living beings. It is of great importance, inasmuch as it provides the appropriate food for the vegetable world, which, under the influence of sunlight, decomposes it into oxygen (which goes to maintain the normal proportion of oxygen in the atmosphere), and at the same time assimilates the carbon.

The actual organic matter in the air is very small indeed—small even in relation to the carbonic acid. There are also very minute traces of nitric acid and of ammonia and other matters.

(3.) *The Atmosphere as a whole, its Extent and Distribution—General Atmosphere above Sea-Level—The Underground Atmosphere.*

The earth's inhabitants live at the bottom of an aërial ocean, which extends upwards for many miles, and penetrates downwards into the soil and the rocks, as well as into the sea itself.

In the plains at the surface of the earth (or more accurately at the level of the sea) the atmosphere has a density so that 773 volumes of air are equal in weight to one volume of water. The average barometric pressure at the level of the sea is equal to a column of 760 millimetres of mercury. As we rise above the level of the sea the barometric pressure diminishes, and the density of the air undergoes corresponding diminution.

By observation it has been found that an ascent of 10.5 metres above sea-level causes a fall of 1 millimetre in the barometer. At 5528 metres above the sea the barometer stands at 380 millimetres, and the air has only half its normal density—386 volumes of air being of equal weight with one volume of water.

Barometric pressure is nothing more nor less than the weight of superincumbent air. The atmosphere existing above the level of the sea is cut in halves at the height of 5528 metres. One-half of the gaseous material lies above the level of the 5528 metres, and the other half of the gaseous material lies between that and sea-level.

The upper regions of the atmosphere are occupied by air of exceeding thinness. How far upwards the atmosphere extends we do not know. It has, however, been computed that the atmosphere existing above sea-level would cover the surface of the globe to the depth of 11 or 12 yards if it were compressed so as to have the density of water, or to the depth of about five miles if it were of uniform density as at sea-level.

As has been pointed out, the atmosphere penetrates downwards below sea-level. There is a large underground atmosphere; there is air in soils and rocks and in the sea. The air of mines, and, to some extent, underground air, has formed the subject of chemical investigation, notably by Pettenkofer and Angus Smith.

The constancy of composition, so remarkable a feature of the general atmosphere above sea-level, does not appertain to underground air at all.

(4.) *General Scope of Sanitary Air-Analysis.*

As has been already mentioned, the general outdoor atmosphere (degree of moisture excluded) is almost absolutely constant in composition all the world over. There is, therefore, very little scope for analysis as applied to the general outdoor atmosphere. Where the air is confined, as in the inside of houses, workshops, cellars, mines, and caverns—in all these instances there is scope, and sometimes very wide scope, for a chemical examination of the enclosed atmosphere.

In mines and caverns we encounter the underground air, which is sometimes so far removed in composition from ordinary air as to be incapable of supporting life or combustion. In cellars and the basement of dwelling-houses there is frequently underground air mixed with the ordinary atmosphere. The air of churches ought to be rigorously examined, for reasons which need not be particularly mentioned.

Defilement of the air is the result of fermentation, putrefaction, respiration, and combustion; and whenever these processes take place, there must be vitiation of the atmosphere, unless there is a free circulation of air. In short, the inside of all buildings affords scope for air-analysis, and the question which air-analysis enables us to answer is this—Is the ventilation of the building sufficient?

CHAPTER I.

EUDIOMETRY AND COMPLEMENTARY EUDIOMETRY.

THERE are two ways of investigating gases. Change in the volume of the gas itself may be made the subject of observation, or the effect on substances exposed to the action of the gas may be recorded.

Thus, for example, we may investigate the composition of a sample of respired air either by noting the diminution of volume occasioned by allowing an alkali to absorb the carbonic acid, or else by weighing the quantity of carbonate of lime which the gas produces when it acts upon lime-water.

Both methods of investigation are valid and logical, and the one is complementary to the other.

The term eudiometry or eudiometrical methods is applied to those processes of gas-analysis which involve measurements of the changes of volume of gases consequent upon analytical operations.

So far as we know, there is no convenient term in use to describe the other kind of gas-analysis, and we are driven to coin a word for the purpose. We propose the term complementary eudiometry. And reverting to the example of respired air—whilst the measurement of the carbonic acid by noting the contraction is eudiometry, the measurement of the carbonic acid by weighing the carbonate of lime becomes a process of complementary eudiometry.

In the hands of Regnault, Bunsen, and other chemists, eudiometry attained to a high degree of precision forty years ago; but the methods of forty years ago involved the employment of considerable quantities of mercury, and were tedious and costly.

The requirements of technical chemistry have called for great simplification of this branch of chemical analysis, and to-day the best modern technical eudiometry is at once accurate and trustworthy, and at the same time rapid and easy of execution, and involves very little outlay of money.

The introduction of the gas-burette, an implement which renders gases almost as manageable as liquids, has greatly aided in the popularisation of gas-analysis. The most convenient of these implements is Hempel's, which we now describe. It consists of two upright glass tubes, one of which is graduated into cubic centimetres, and the other plain. The graduated tube is narrowed almost to capillarity at the top, and drawn out so as to take an india-rubber connecting tube. Both tubes are fitted into stands at the bottom, and provided with side-tubes. About a yard of wide india-rubber tube connects these bottom side-tubes. The graduated upright tube is designed for the reception and measurement of the gas. The plain upright tube is the pressure-tube, and its function is to regulate the pressure in the other tube. *Vide* Plate I.

In order to use the gas-burette, water must first be introduced, so as to rather more than half fill it. By raising the pressure-tube the water will be caused to fill the graduated tube. By lowering the pressure-tube a sample of air or gas may thus be drawn into the graduated tube. A pinch-cock closes the india-rubber connecting tube, and confines the sample of gas in the graduated tube.

The next operation is to read off the volume of the gas. For this purpose it must be placed under the current barometric pressure by raising or lowering the pressure-tube until the level of the water is the same in both tubes. Then the graduation must be read off.

Having been measured, the sample of air or gas is submitted to a variety of analytical operations in gas-pipettes, and afterwards measured in the gas-burette in the same manner as before.

The annexed drawing of Hempel's absorption-pipette will, to some extent, explain itself. The bulbs are of considerable size, the large bulb, for solution of caustic potash, having a capacity of about 180 cubic centimetres. Several absorption-pipettes are required, and each particular absorption-pipette is appropriated to its own use.

By means of india-rubber connecting tubes and bent capillary glass tubes the gas-burette is placed in communication with the absorption-pipettes, which require to be raised and supported on blocks of wood. The india-rubber connections should be bound with thin copper wire. The connection and disconnection of the absorption-pipettes with the gas-burette requires a little practice in order to do it properly. The capillary tube being about one millimetre in diameter, it follows that a metre of the tube will hold about three-quarters cubic centimetre of gas. There should not be more than one-tenth of a cubic centimetre of space between the gas-burette and the absorption-pipette.

In using an absorption-pipette, the liquid should never be allowed to touch the india-rubber connection, but should be stopped at a mark placed on the enamel background. Each absorption-pipette requires its own pinch-cock, its own india-rubber connector, and its own glass tube bent at a right angle (*vide* fig. 1). A common india-rubber connector, *c*, binds together the bent glass tube of the gas-burette with the bent glass tube of the absorption-pipette. The connector *c* (which has frequently to be altered) should be wiped dry, and the ends of the glass tubes should be wiped dry and brought close together within the india-rubber connector *c*, which must be properly bound. Before altering connector *c*, both pinch-cocks should be made to close the connections with the burette and the pipette.

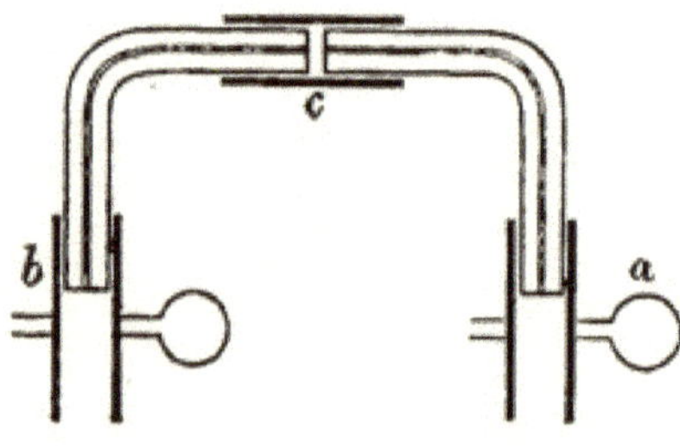

FIG. 1.

It will be readily understood that the sharpness of the results depends upon maintaining the gas at the same temperature and pressure when its volume is read off. The gas-burette should not be touched with the hand, but may be moved by laying hold of the foot into which the bottom is fitted. The importance of quick processes is likewise manifest.

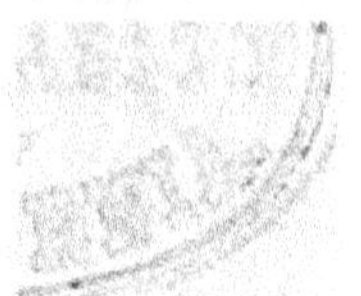

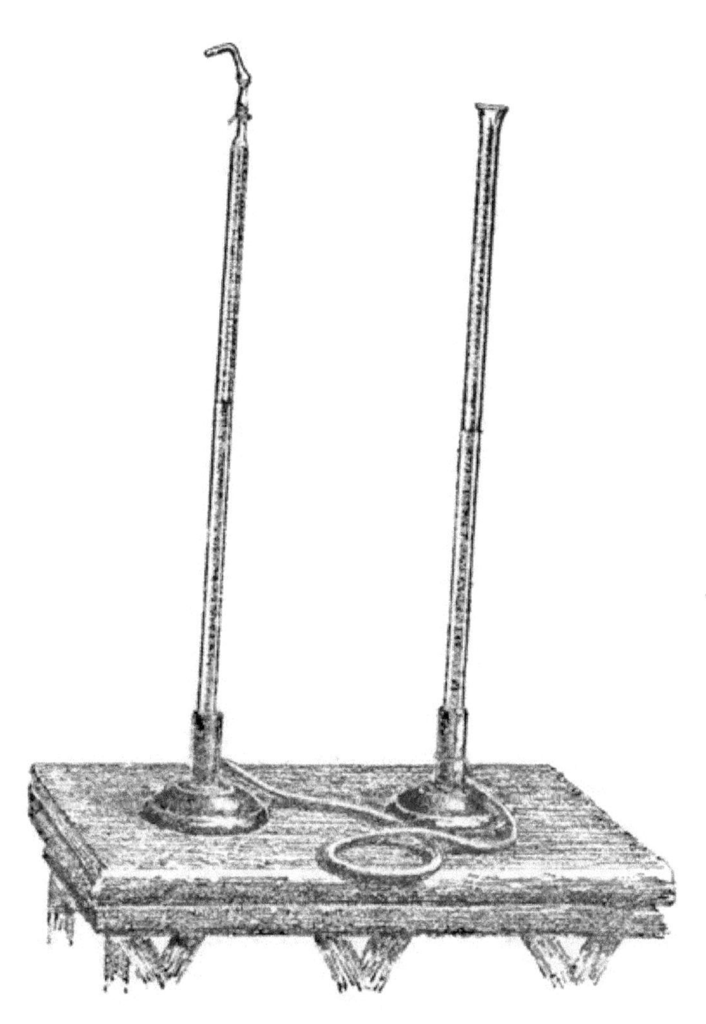

Hempel's Gas-Burette, Ready for Use.

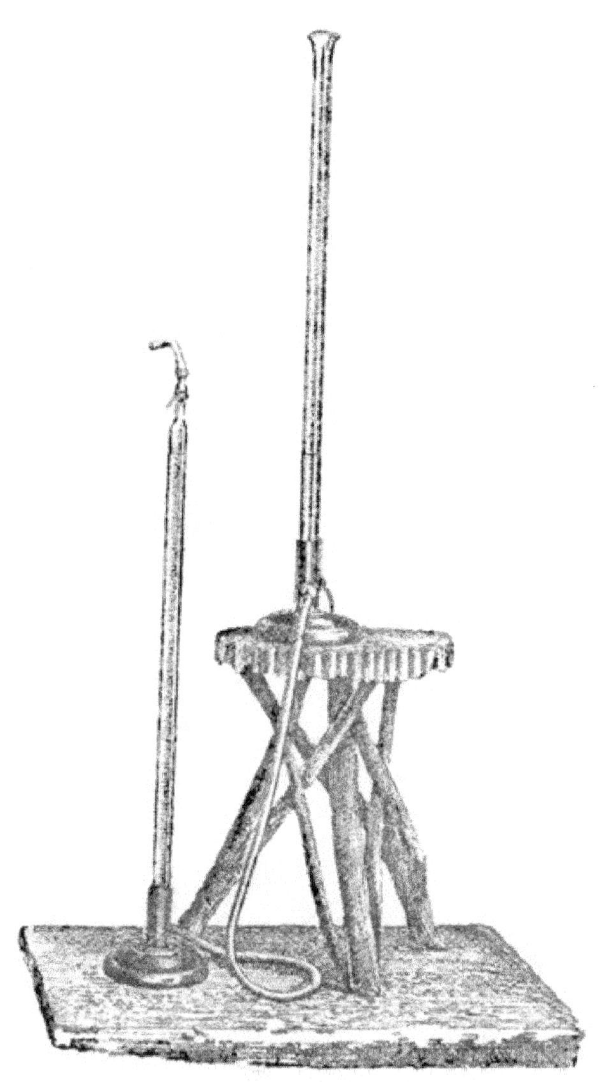

PRESSURE-TUBE RAISED.

PLATE III.

PRESSURE-TUBE LOWERED.

PLATE IV.

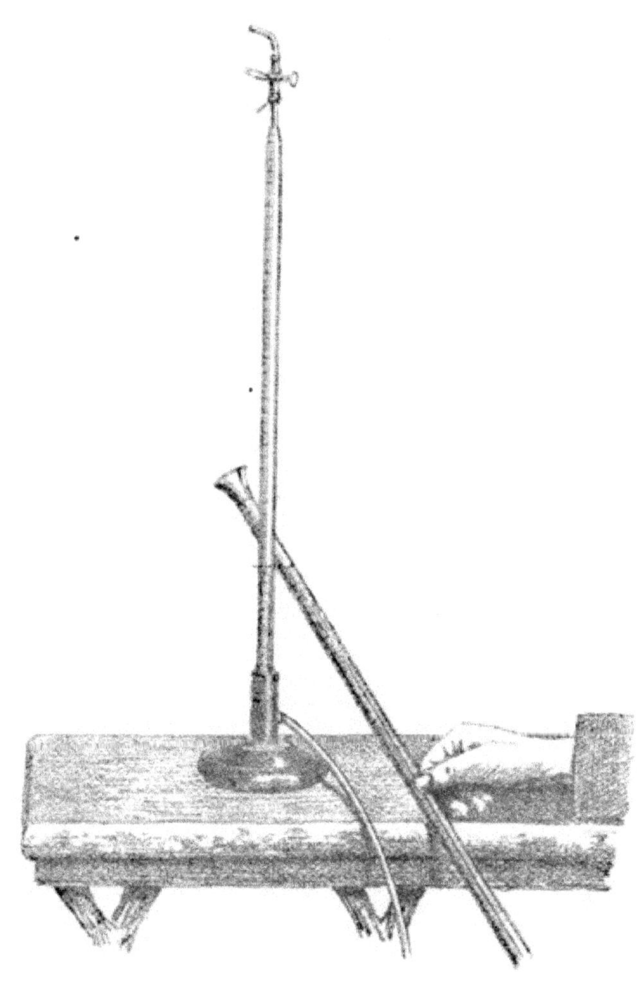

A Reading.

PLATE V.

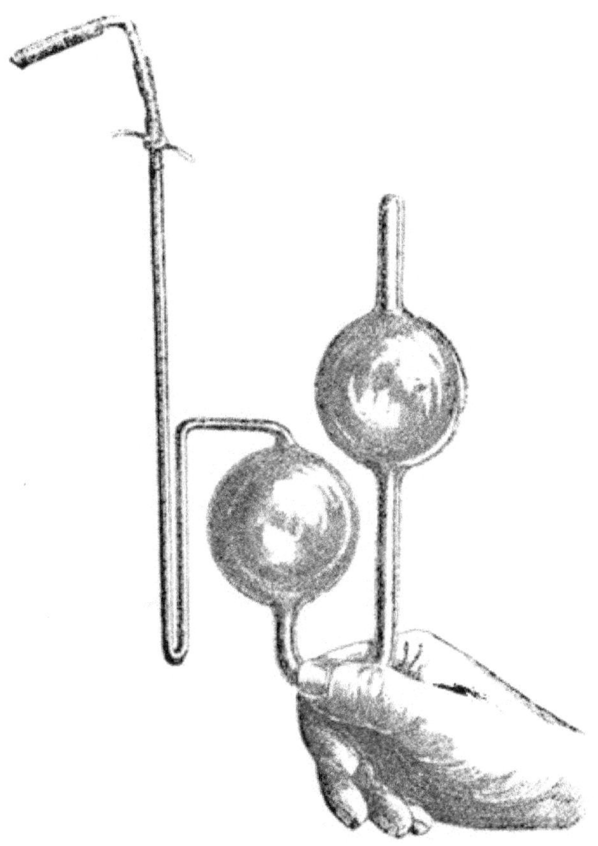

The Absorption-Pipette.

PLATE VI.

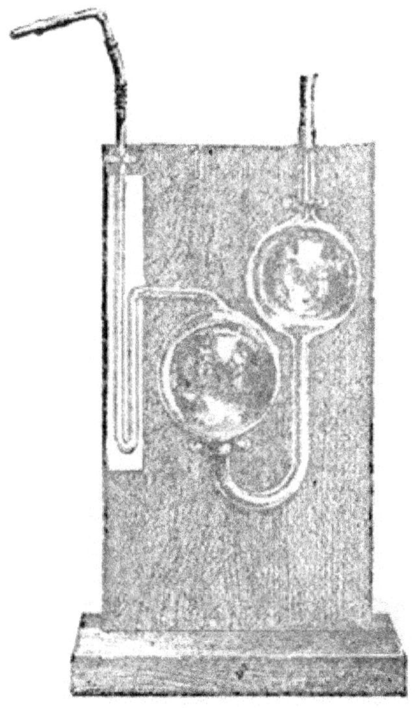

ABSORPTION-PIPETTE MOUNTED.

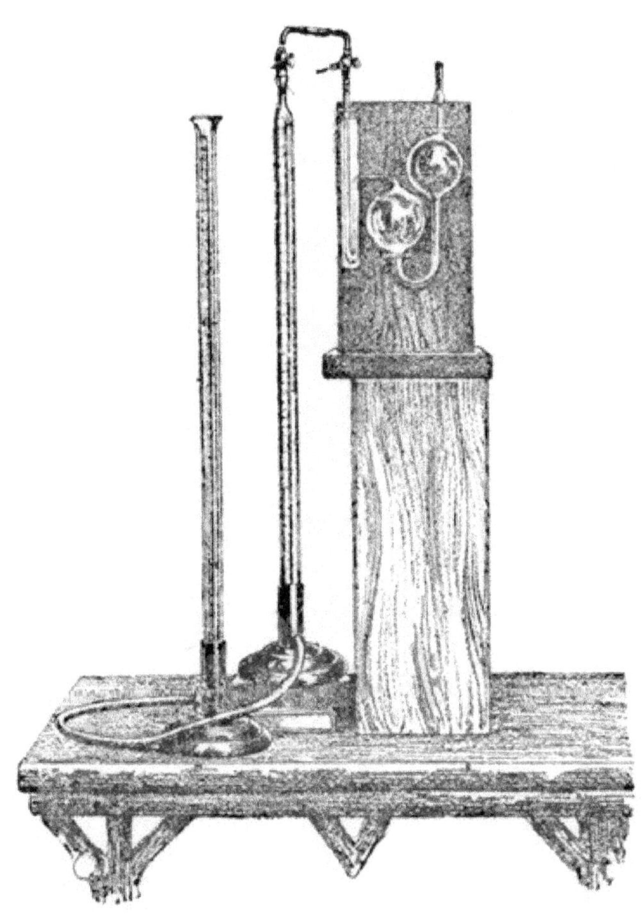

Connected and in Use

LIBRARY
OF THE
UNIVERSITY
OF
CALIFORNIA.

CHAPTER II.

OXYGEN (O_2).

PRIESTLEY and Cavendish employed nitric oxide as a means of measuring the proportion of oxygen in the atmosphere. The method, however, fell into disrepute early in the present century. We quote the following passage from Watts's "Dictionary of Chemistry" (1866), vol. iv. p. 68:—

"Nitric oxide was formerly used, especially by Priestley and Cavendish, to estimate the proportion of oxygen in the air or other gaseous mixture; but the method is not capable of yielding exact results on account of the difficulty of obtaining nitric oxide perfectly pure, and it has long since been superseded by more trustworthy methods."

We are now able to restore this old method to its proper place; in our hands it has proved very accurate.

The reaction between oxygen and nitric oxide is—

$$2NO + O_2 = 2NO_2.$$

The NO_2 is absorbed by water. There is, therefore, a contraction of three volumes for every one volume of oxygen. The mode of operation is to add excess of nitric oxide to the sample of air, and then to read the contraction. One-third of the contraction is the volume of the oxygen in the sample. By operating in that manner we avoid all difficulty arising from the presence of impurity in the nitric oxide.

The following examples will serve to show the accuracy of the process :—

	I.	II.	III.
Vol. of air taken . .	80.0	50.0	70.0
„ NO added . .	45.0	25.0	54.0
	125.0	75.0	124.0
Vol. after reaction	75.6	44.2	80.6
Therefore, contraction . . .	49.4	30.8	43.4
„ oxygen	16.47	10.27	14.47
Percentage of oxygen	20.59	20.54	20.67

The air is measured in the gas-burette and passed into an absorption-pipette charged with water. Nitric oxide is introduced into the gas-burette and its volume measured, and then it is passed to the air in the absorption-pipette. Instantly the well-known reaction takes place, and ruddy fumes of NO_2 make their appearance in the bulb of the absorption-pipette. The absorption of the ruddy fumes by the water is very quick. The gas is passed backwards and forwards once or twice—the fumes have vanished—and the final reading may be at once made. Nothing is gained by the substitution of caustic alkali for water in the absorption-pipette.

The nitric oxide is prepared by the action of dilute nitric acid on copper-turnings, and one preparation of the gas will suffice for many analyses. The gas is preserved in a bell-jar over water.

Oxygen may also be advantageously measured by means of pyrogallic acid in presence of alkali. The absorption-pipette is charged with a solution containing about 15 grammes of pyrogallic acid and 50 grammes of solid caustic potash. The absorption of the oxygen is by no means instantaneous, but it is rapid. In order to be sure of the result, it is absolutely requisite to repeat the readings until they are constant. Once charged, as above recommended, the absorption-pipette will suffice for a great number of measurements of oxygen.

Our own measurements of oxygen in air by this process have given—

	I.	II.	III.
Percentage of oxygen . .	20.0	20.8	20.5

Explosion with Excess of Hydrogen.—Of all the methods used to measure the oxygen in air, this is the one which is most esteemed by chemists. Operating over mercury, and taking every precaution, Bunsen and Regnault, in years gone by, obtained results wonderfully constant and concordant.

Hempel's new gas-apparatus includes a bulb-apparatus provided with platinum wire, so as to admit of explosion by means of the electric spark. There is also a hydrogen-pipette in which hydrogen is generated and stored for use. We have used the apparatus, and can testify to the practicability of the arrangement. Everything being in order, a measurement of the oxygen by explosion with hydrogen can be got through in about a quarter of an hour.

The steps of the operation are as follows:—

The air is measured in the gas-burette; hydrogen (about half as much hydrogen as air) is then added to the air, and the total volume read off. The gas-burette is connected with the explosion-bulbs, and the air and hydrogen passed into them. The electric spark is then passed, and the explosion takes place. The gas is then passed back into the gas-burette and its volume read off. One-third of the contraction equals the volume of the oxygen. The circumstance that the explosion occurs in a separate apparatus, and not in the gas-burette itself, renders it possible to make a reading very soon after the explosion. The weak point in explosion processes in Hempel's apparatus is the danger of some escape of gas through the joints; and these measurements of oxygen in air in the Hempel apparatus are liable to be slightly in excess. Our results are:—

	I.	II.	III.
Percentage of oxygen in air .	21.34	20.94	21.34

The slight excess in Experiments I. and III. we believe arises from leakage through the joints (leakage would cause undue contraction).

Of the three methods for the measurement of oxygen,

we prefer the first for many reasons. It is more widely applicable than either of the others.

Neither the pyrogallic acid process nor combustion with excess of hydrogen is available in presence of carbonic acid. The nitric oxide method, on the other hand, is unaffected by the presence of carbonic acid.

CHAPTER III.

NITROGEN (N_2).

NITROGEN is recognised by its inertness. It is not absorbed by alkalies. It is attacked neither by oxidising nor by reducing agents. It is the residue left when air is exposed to the action of pyrogallic acid in presence of potash. It is left also when air is passed through a tube containing red-hot metallic copper.

A mixture of nitrogen with hydrogen, or with carbonic oxide, or with a hydrocarbon gas—marsh-gas, for instance—may be recognised by its property of leaving a residual gas insoluble in potash when it is burnt by being passed through a glass tube charged with red-hot oxide of copper.

In addition to oxide of copper, there should be some metallic copper in the tube, and operating in that manner, a "Dumas" determination of nitrogen is accomplished.

The broad general statement, that nitrogen neither burns itself nor is a supporter of combustion, is not absolutely true under all possible conditions. Mixed with other combustibles, and with excess of oxygen, it combines freely with oxygen at a very elevated temperature, and there is a little oxidation of nitrogen in common combustions. The following quotation from Bunsen's "Gasometrische Methoden" (second edition, page 71) is to the point:—

"We can very easily combine nitrogen direct with oxygen, thereby forming nitric acid, when we explode a mixture of these gases with twice its volume of detonating gas (Knallgas). If the detonating gas reaches three times or five times the volume, then there is so much nitric acid produced that mercury dissolves and yields crystals of nitrate of mercurous oxide."

By detonating gas (Knallgas) Bunsen means the mixture of two volumes of hydrogen with exactly one volume of oxygen, obtained by the electrolysis of water.

The oxidation of the nitrogen in these cases depends upon the high temperature which is attained when the proportion of detonating gas is large. On the other hand, when the temperature is kept comparatively low, and when the proportion of detonating gas is small, no oxidation of nitrogen takes place.

The following experiments have been published by Bunsen:—

100 volumes of air mixed with 13.45 volumes of detonating gas do not explode with the electric spark.

100 volumes of air with 26.26 c. c. of detonating gas are capable of explosion with the electric spark,

and no residue of detonating gas remains unexploded after the explosion has taken place.

Gradually increasing the proportion of detonating gas, Bunsen found that the 100 volumes of air remained quite constant after explosion until 64.31 volumes of detonating gas had been added. With more than that proportion of detonating gas there is oxidation of the nitrogen of the 100 volumes of air.

Bunsen lays down the rule that in order to attain accuracy in gas-analysis by explosion, the proportion of combustible mixture must not be larger than 64, nor smaller than 26, to 100 volumes of inactive gas.

LIBRARY
OF THE
UNIVERSITY
OF
CALIFORNIA.

ABSORPTION-PIPETTE CHARGED WITH POTASH.

CHAPTER IV.

CARBONIC ACID (CO_2).

Carbonic acid combines with, and is readily absorbed by, caustic alkalies. In order to measure the quantity of carbonic acid in a gaseous mixture containing a considerable proportion of carbonic acid, the following method may be resorted to, and Hempel's apparatus may be used.

The *absorption-pipette* is charged with a solution made by dissolving one part of solid sticks of potash in three parts (by weight) of water. A funnel with the neck properly narrowed will be found useful in getting the solution into the absorption-pipette. Sufficient liquid should be poured in, so that the large bulb is quite full, and at the same time there should be an empty space of at least 100 cubic centimetres in the other bulb. An ink-mark should be made on the enamel at a convenient spot, and to this line the solution should be raised. A wooden support (a box, or a light stool, or other convenient form) raises the absorption-pipette and allows of easy connection with the gas-burette.

The sample of gas to be analysed is measured (as

has been described in Chapter I); connection is next made with the absorption-pipette, and the gas forced over into the pipette. About a minute is allowed to elapse, then the gas is drawn back into the gas-burette, care being taken to stop the potash exactly (or nearly exactly) at the ink-mark on the enamel. The volume of gas is read off.

The operation is repeated, that is to say, the gas is forced into the potash bulbs and drawn back into the gas-pipette, and the volume of gas again read off. If the reading be the same as before (or not more than one-fourth cubic centimetre different), the experiment is at an end, and the result may be recorded in the note-book. The potash bulbs may then be disconnected, and the residual gas is available for further examination.

Ordinary atmospheric air, when subjected to the treatment just described, exhibits no sensible amount of carbonic acid, because the normal amount of that gas in the air is less than one-tenth per cent. Neither with Hempel's apparatus, nor with mercury in the manner recommended by Bunsen or Regnault, is it possible to measure the minute proportion of carbonic acid in the atmosphere. This is one of those cases where eudiometry fails, and where complementary eudiometry comes in.

Respired air, on the other hand, contains amply sufficient carbonic acid to be dealt with by eudiometry. We have obtained the following results with different samples of respired air, the Hempel's apparatus having been used as above described.

In 100 volumes of respired air we found—

Example I.	4.5 vols. of CO_2	
,, II.	4.2 ,,	,,
,, III.	4.4 ,,	,,
,, IV.	5.0 ,,	,,

Respired air is very easily procured for these experiments, all that need be done being to breathe three times through a narrow glass tube into a Winchester quart bottle, which in that manner becomes charged with respired air. The bottle may be stoppered and the respired air employed at leisure, a narrow glass tube in connection with the gas-burette being used to draw it from near the bottom of the Winchester quart bottle when wanted for experiment.

By holding the breath, or by a second time breathing respired air, even a higher percentage of carbonic acid may be shown.

In mines the air sometimes exhibits very high percentages of carbonic acid, but badly ventilated rooms do not exhibit anything of this sort. As has already been mentioned, the proportion of carbonic acid in the general atmosphere is far too minute to admit of being measured by eudiometrical processes, and the other method, viz., complementary eudiometry, has to be adopted.

The principle of the method is the measurement of carbonic acid by the quantity of carbonate of lime, or by the quantity of lime which it is able to saturate. Early in the century De Saussure published an investigation of air in Switzerland, and employed the method of weighing the carbonate of lime. Later on, Pettenkofer of Munich brought out the process which is now almost universally used in such investigations. We use Pettenkofer's process.

In measuring the carbonic acid in the air, it is essential to operate upon a large volume. In our laboratory we take a large stoppered bottle of colourless glass, holding rather more than a gallon. The capacity of the bottle was accurately measured by measuring the volume of the water required to fill the bottle. The volume was found to be 5350 cubic centimetres. In using the bottle, means have to be taken to ensure the entrance of the desired sample of air into the bottle. For this purpose we have a very simple kind of aspirator. The aspirator consists of two round boards, one of which is perforated to allow of the entrance or exit of air through a narrow pipe, and the boards are connected by sheet india-rubber, mackintosh, or leather (*vide* fig. 2).

FIG. 2.

The arrangement forms a sort of bellows. When the boards are drawn asunder, air is sucked into the bellows, and when they are pressed together, air is forced out.

When we desire to charge the bottle with the air of a room, we bring the bottle into the room, take out its stopper, and suck out the air from the bottle by means of the aspirator, the pipe of the aspirator being first inserted into the bottle, until the aspirator is fully expanded, and then removed from the bottle to allow of the extracted air being got rid of by pressing the boards together. Our habit is to extract the air four times, so as to make sure that the bottle becomes properly charged with the desired sample of air. As soon as the bottle is charged, the stopper is put into its place and screwed round, so as to secure the charge of air.

The sample of air having been properly secured, the next step is the measurement of the volume of carbonic acid contained by it. Lime-water has the property of combining with carbonic acid, even when the carbonic acid is exceedingly diluted with air, and in combining with carbonic acid, the lime-water becomes exhausted, more or less completely, depending upon the amount of carbonic acid existing in the sample.

In order to measure the carbonic acid, we require lime-water (the strength of which we can ascertain by means of standard acid), and we require a suitable standard acid. We find it convenient (as has been recommended by chemists who have cultivated this branch of chemical analysis) to make a special standard acid for this purpose. This acid, which we term

Liquid Substitute for Carbonic Acid Gas, is of such a strength that it neutralises just as much lime as its own volume of carbonic acid gas would neutralise. The volume of carbonic acid is taken at 15° C. and 760 millimetres barometric pressure. At that temperature and pressure one cubic centimetre of carbonic acid gas weighs 1.86 milligrammes, and one cubic centimetre of the acid is equivalent to 1.86 milligrammes of carbonic acid.

Such acid is made by diluting what is known as *Normal* acid,* so that 10 volumes of normal acid are expanded to 118 volumes.

On reflection, our readers will perceive how convenient the acid is. It is the representative of the carbonic acid in the sample of air.

We take a given volume of lime-water and find how many cubic centimetres of "Liquid Substitute for CO_2 Gas" are required to neutralise it. We then take the same volume of lime-water and pour it into the bottle of air and absorb the carbonic acid. When the absorption is finished, viz., in from half an hour to an hour, we add one cubic centimetre of solution of litmus, and then measure in the *liquid substitute*, noting carefully the point at which the blue colour changes to violet, and in that way read off the quantity of liquid substitute required by the lime. The difference is the volume of carbonic acid existing in the sample of air.

* Normal acid contains an equivalent of acid in milligrammes in one cubic centimetre.

The following examples will serve in explanation of the method :—

Experiment I.—The bottle was charged with air from a garden in New Malden. The volume of lime-water was 25 c. c., which neutralised 11.6 c. c. of "liquid substitute." After absorption of the carbonic acid in the sample of air, the lime in the bottle required 9.5 c. c. of liquid substitute. Therefore the volume of carbonic acid gas was 2.1 c. c., and the percentage of CO_2 in the sample of air was 0.040.

Experiment II.—Air from a large well-ventilated dwelling-room—

Volume before absorption . .	11.5 c. c.
,, after ,, . .	9.2 c. c.
	2.3 c. c.

Percentage of $CO_2 = 0.043$.

Experiment III.—Air from small sitting-room—

Volume before absorption . .	11.5 c. c.
,, after ,, . .	8.5 c. c.
	3.0 c. c.

Percentage of $CO_2 = 0.057$.

Experiment IV.—Air from room with two gas-lights and three people—

Volume before absorption . .	11.0 c. c.
,, after ,, . .	6.5 c. c.
	4.5 c. c.

Percentage of $CO_2 = 0.085$.

It is not necessary to remind our readers that work of this sort is delicate, and requires cleanliness and precision; and indeed it is desirable to increase the volume of air taken for analysis. Recent experiments of our own have convinced us that this is quite practicable. Instead of being content with one bottleful of air, we take several bottlefuls of air, using, however, only one charge of lime-water. We find that an exposure of ten minutes to the action of the lime-water is sufficient for the absorption of the carbonic acid from one bottleful of air, if the precaution be taken of causing the bottle to revolve sixty times, so as continually to bring fresh surface of lime-water into contact with the gas. A ready way of doing this is to roll the bottle across the floor of the laboratory for a certain number of times. Or the bottle might be placed in a frame and made to rotate by any convenient mechanical appliance.

The following examples will serve in illustration. Air from a garden in New Malden, Surrey, was examined three times on the same day.

Experiment I.:—

Capacity of bottle . . .	5620 c. c.
Volume of lime-water . . .	50 c. c.
	5570 c. c.
Number of times of filling .	4
Volume of air taken . .	22,280 c. c.

The 50 c. c. of lime-water was capable of saturating

21.4 c. c. of liquid substitute before being used, and after being used it saturated 14.8 c. c. Therefore 6.6 c. c. of CO_2 gas was contained by the 22,280 c. c. of air, or, 100 volumes of the air contained 0.0296 volumes of CO_2.

Experiment II. gave that 10.4 c. c. of CO_2 was contained by 31,800 c. c. of air, or $CO_2 = 0.0327$ per cent.

Experiment III. gave that 8.4 c. c. of CO_2 was contained by 27,850 c. c. of air, or $CO_2 = 0.0302$.

The date was 29th April 1890; barometer 755 millimetres; temperature 14.5° centigrade.

The usual directions in the text-books run to the effect that the air-bottle in these experiments should be carefully cleaned and dried before each experiment. To do so involves a degree of trouble and loss of time which we find to be quite uncalled for. On the conclusion of each testing of a sample of air, the bottle contains spent and neutralised lime-water. All that need be done is to pour this liquid out of the bottle, which then becomes fit for the reception of the next specimen of air. No improvement whatever (but rather the reverse) is brought about by washing and drying in such a case.

It has been said that the only appropriate acid to use for this testing is oxalic acid, and that turmeric ought to be used instead of litmus. We do not concur. Our experience is that the lime-water behaves admirably with sulphuric acid or hydrochloric acid, using litmus as indicator. The strength of the 50 c. c. of lime-water we have no difficulty in measuring within 0.1 c. c. of the "liquid substitute." And after the partial saturation with carbonic acid from the air, we have no difficulty in reading sharply. The point to read to is the production of the violet tint—not the full red.

CHAPTER V.

RESPIRED AIR—AIR FROM FURNACES—CARBONIC OXIDE, ITS MEASUREMENT IN AIR—PHYSIOLOGICAL ASPECT OF THE SUBJECT—RULES FOR REPORTING ON AIR.

THE analysis of respired air offers points of interest to which we desire to direct attention. The following analyses were made by ourselves on 24th March 1890. The air was breathed into a Winchester quart bottle, the breath being held a while, so as to obtain highly spent air. Hempel's apparatus was used.

	I.	II.
Vol. of air taken	66.8	99.0
Vol. after absorption of CO_2 .	63.4	94.0
Vol. of NO added . . .	45.0	60.0
	108.4	154.2
Vol. after reaction	80.5	108.6
Therefore contraction . .	27.9	45.6
Therefore oxygen	9.3	15.2

which gives in percentage:—

	I.	II.
Carbonic acid	5.1	5.0
Oxygen	14.0	15.4
Nitrogen	80.9	79.6
	100.0	100.0

These analyses illustrate very well the way in which the living animal alters the composition of the air during the process of respiration. They are extreme cases, but they show the well-ascertained fact that the withdrawal of oxygen is rather greater than the corresponding evolution of carbonic acid.

By the combustion of fuel the oxygen of the air is diminished and carbonic acid evolved. A very common form, and a very dangerous kind of combustion, is that in which the combustion is incomplete, so as to yield carbonic oxide. We have recently analysed a specimen of air escaping from a furnace as follows:—

Carbonic acid	6.0
Carbonic oxide	20.0
Residual nitrogen, &c. . . .	74.0
	100.0

Such products of combustion escaping into a room are very dangerous to health.

The history of the carbonic oxide is this:—The furnace contains a thick layer of red-hot coke, air enters at the bottom of the furnace, becomes carbonic acid, which undergoes reduction so as to yield carbonic oxide in passing up through the furnace—

$$CO_2 + C = 2\,CO.$$

This chemical change likewise takes place in common fires when the fuel is burning low, and owing to the imperfections in the action of the chimney, carbonic oxide often finds its way into the air of dwelling-rooms. Carbonic oxide is notoriously poisonous and injurious to health, and the air not only of factories, but of dwelling-houses as well, ought to be examined for carbonic oxide.

In the above analysis the carbonic oxide was measured by noting the volume absorbed by solution of subchloride of copper in the Hempel apparatus.

The subchloride of copper is prepared by digesting oxide of copper and copper turnings in strong hydrochloric acid. Either the solution in strong hydrochloric acid may be used at once (as in the above experiment), or else the subchloride may be precipitated by the addition of water, and the resulting precipitate dissolved in ammonia, and the ammoniacal solution placed in the Hempel bulbs and used. The absorption of carbonic oxide by subchloride of copper is not quite so quick as the absorption of carbonic acid by potash; and the operator is warned to adopt precautions so as to avoid an incomplete absorption. That is to say, the operator must repeatedly pass back the gas into the absorption-pipette, and then send over the gas and re-read in the gas-burette. Only when there is no difference between two consecutive readings can the operator be sure that all carbonic oxide has been removed.

The further caution must be given that, before attempting the measurement of the carbonic oxide in a sample of gas, the operator must make sure that oxygen is absent.

Obviously, also, this method can only answer where the amount of carbonic oxide is considerable. Where traces are concerned—just as in the instance of carbonic acid—another method must be resorted to.

The method which we propose is as follows :—

We take a large volume of air, extract the carbonic oxide by means of solution of subchloride of copper in hydrochloric acid, and then place the copper solution in a suitable vessel and drive out the dissolved carbonic oxide by means of acid solution of bichromate of potash, and collect and measure the evolved carbonic oxide.

This mode of dealing with traces of carbonic oxide was described by one of us in the *Gas World* some time ago.

In the special instance of the presence of traces of carbonic oxide in atmospheric air, the oxygen is a hindrance, because it uses up the copper solution. We get over the difficulty by adding nitric oxide to the air and absorbing the resulting NO_2 by means of water. Should there be any attack of the carbonic oxide by the nitric peroxide, we may employ a strongly alkaline solution with protoxide of manganese in suspension, which absorbs the oxygen pretty completely in about twenty-four hours.

The practical details are these. The sample of air being contained by four air-bottles (such as we use for the examinations of the CO_2 in the air), we first remove the major portion of the oxygen, as just described, and then pour 50 c. c. of acid subchloride of copper solution into one of the bottles, and roll the bottle, or rotate it as if for CO_2, only for a much longer time. When it has exhausted the first air-bottle of carbonic oxide, the 50 c. c. of copper solution is poured into the second air-bottle, and so on till all four air-bottles are disposed of. Finally, the 50 c. c. of copper solution (which now contains the carbonic oxide from the four air-bottles) is placed in an appropriate vessel and treated with acid solution of bichromate of potash, and the evolved carbonic oxide collected, measured, and investigated in any way that may seem desirable (*vide* Plate IX.).

We cannot give any examples at present, but we believe that an investigation of the air in workshops and dwelling-rooms would often disclose the presence of carbonic oxide in measurable quantity.

At the beginning of the chapter we cited the analysis of air from the recesses of the lungs; we have also set out the composition of the gaseous products from the furnace. It is, so to speak, the sewage of the atmosphere. Air fit to breathe, far from being anything of the sort, ought not to contain one per cent. of these defilements.

Taking the normal carbonic acid in the air as 0.035 per cent., the addition of one per cent. of respired air would raise the carbonic acid to 0.085 per cent., which, on that reckoning, indicates bad air.

Viewing the question from a physiological standpoint, we place before our readers the following considerations.

In twenty-four hours the average quantity of carbonic

acid gas evolved by the lungs of a full-grown man may be set down as half a *cubic metre*, that is to say, 500 litres; and the total volume of air breathed in twenty-four hours is about 13 cubic metres, or 13,000 litres. Between 30 and 40 lbs. of air are breathed daily by each of us, and it is hardly an exaggeration to say that our aërial food which enters by way of the lungs dwarfs our liquid and solid food into insignificance.

Minute changes in the percentage composition of the air taken into the lungs express quantities of material which mount up to large absolute weights in the course of the day. So small a percentage of carbonic acid as 0.01 amounts to 1300 cubic centimetres of carbonic acid on the day's consumption of air, and weighs 2.4 grammes (or 37 grains). Absorption through the lungs is notoriously most efficient, and if the 37 grains of CO_2 were replaced by 37 grains of actively poisonous organic matter, it would be enough to be fatal to thirty persons. There is nothing far-fetched or unreasonable in attaching importance to minute changes in the composition of the air we breathe.

The importance of taking note of the carbonic acid in the air has been generally recognised. A large proportion is injurious, because carbonic acid *per se* is incapable of oxygenating the blood, whilst very small increments over and above the normal figure in the atmosphere are to be objected to as representatives or concomitants of organic matter of an injurious kind. So far as we know, there is no way in daily life in which the carbonic acid in the air is sensibly increased without damage being done to the air.

The following may be laid down :—

Normal air out of doors contains 0.03 to 0.04 volumes of CO_2 in 100 volumes of air.

Air indoors, in properly ventilated and not over-crowded rooms, should not contain more than 0.06 volumes of CO_2 per cent.

Air which contains 0.08 volumes of CO_2 per cent. is contaminated.

Air which contains 0.10 volumes of CO_2 per cent. is distinctly too foul, and the figure 0.20 represents something very bad indeed.

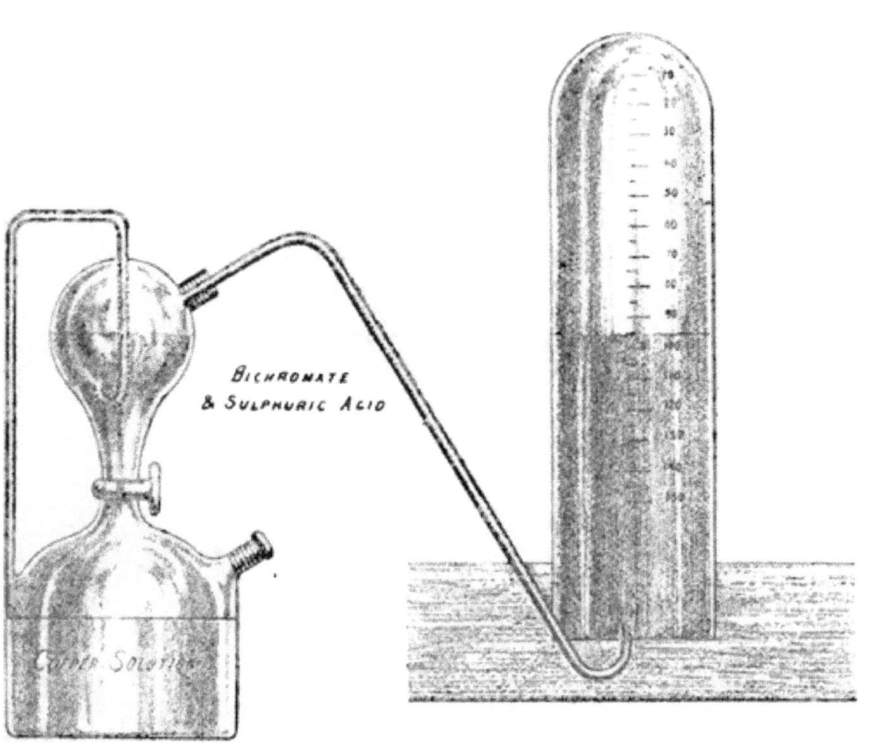

Recovery of Carbonic Oxide Gas.

LIBRARY
OF THE
UNIVERSITY
OF
CALIFORNIA.

APPENDIX.

ON ILLUMINATING GAS.

BUNSEN'S analysis of coal-gas, as supplied in Heidelberg, is as follows:—

	Volume.	Weight.
H_2, Hydrogen	46.20	6.97
CH_4, Marsh-gas	34.02	41.09
CO, Carbonic oxide	8.88	18.77
C_6H_6, Benzene	1.33	7.83
C_3H_6, Propylene	1.21	3.84
C_2H_4, Ethylene	2.55	5.39
CO_2, Carbonic acid	3.01	10.00
N_2, Nitrogen	2.15	4.54
O_2, Oxygen	0.65	1.57
	100.00	100.00

The hydrogen and marsh-gas and carbonic oxide (which make up 89 out of the 100 volumes) are non-illuminants, and only indirectly contribute to the illuminating power by the heat evolved by their combustion. The illuminants are the benzene, propylene, and minute traces of naphthalene and other complex organic compounds.

A very small proportion of the entire bulk of coal-gas consists of illuminants, which are carbon compounds of large molecular weight, whilst the great bulk of the gas is simply heating gas.

As will be seen on examining the analysis, about one-third of the volume of the gas consists of marsh-gas; and on making a calculation it will be found that the ratio of carbon to hydrogen in the gas, taken as a whole, is the same as in marsh-gas—viz., 3 : 1.

Owing to the large proportion of hydrogen, the specific gravity of the gas is low—viz., 0.4584 (air = 1.00).

Bunsen's analysis, carried out with every precaution and in the mercurial trough, gave—

Volume of gas taken . . .	1.000
Contraction	1.532
Carbonic acid	0.569
Air consumed	5.58

In the year 1888 we examined the gas supplied to the town of Stockport in Cheshire, and worked with the Hempel apparatus. Our results, which we believe to be fairly accurate, show the practicability of performing analyses involving explosion. Fair accuracy, though, as we have mentioned, not the highest degree of accuracy, is attainable in that manner.

The following are the details :—

Vols. at Current Pressure and Temperature on 31st May 1888.	Sample I.	Sample II.	Sample III.
Vol. taken	9.0	9.6	8.2
+ Air	99.0	100.0	87.4
After explosion	85.4	85.0	74.4
After absorbing CO_2 . . .	80.4	79.8	69.8
After pyrogallic and potash * .	73.4	74.1	63.8

Therefore :—

	I.	II.	III.
Vol. of gas	1.00	1.00	1.00
Contraction .	1.51	1.56	1.59
Carbonic acid .	0.555	0.54	0.56
Air consumed .	6.30	6.59	6.17

* A point, which is not sufficiently insisted on, is brought out by these analyses. In each instance the end-nitrogen left after absorption of oxygen by pyrogallic acid is greater than the nitrogen added in the shape of air, viz. :—

	I.	II.	III.
End-nitrogen	73.4	74.1	63.80
Nitrogen added as air . . .	71.14	72.24	62.60
	2.26	1.86	1.20

These figures, however, do not represent the nitrogen in the Stockport gas, but are errors of experiment. The end-determination of nitrogen in analysis involving explosion in Hempel's apparatus is not very accurate, because it is affected by the accumulated experimental error arising in the whole course of the manipulation. The explosion in the bulbs charged with water is calculated to extract traces of nitrogen from solution in the water; and, moreover, by the time the end-nitrogen is arrived at, the gas has been three times in contact with considerable volumes of liquid in the absorption-pipette. The circumstance that a comparatively small quantity of gas is taken for explosion—in the present instance only 9 c. c.—renders the matter worse. No confidence can be placed in determinations of nitrogen in these cases. When the nitrogen has to be measured, the best way is to resort to a proper Dumas' nitrogen determination.

On 31st August 1888, the Stockport gas was further investigated, viz. :—

Volume taken	96.0
After potash and pyrogallate . .	94.8
After Cu_2Cl_2	90.2

Therefore, in 100 volumes of Stockport gas :—

Carbonic acid and trace of oxygen .	1.25
Carbonic oxide	5.00
	93.75
	100.00

Comparing this with Heidelberg gas, as examined by Bunsen, we have 1.25 of carbonic acid and oxygen as against 3.66. Stockport gas should consequently require rather more air for total combustion than Heidelberg gas does. We expect that the slightly larger consumption of air, as shown by our analysis, is not altogether due to error of experiment.

The datum, how much air is used up for the complete consumption of coal-gas, is often asked for, and the answer is: One volume of gas takes from 5.5 to 6.5 volumes of air.

Coal-gas admits of being analysed (without explosion) as follows :—

First, The illuminants may be absorbed by strong alcohol in an absorption-pipette, the vapour of the alcohol being of course removed by the water in the gas-burette.

Second, The CO_2 is absorbed by potash in the absorption-pipette.

Third, Traces of oxygen are absorbed by pyrogallate of potash.

Fourth, CO is absorbed by solution of subchloride of copper either in hydrochloric acid or ammonia.

Fifth, Hydrogen gas is absorbed by alkaline solution of permanganate of potash. The residue is marsh-gas and nitrogen, which are not touched by alkaline permanganate even at 212° Fahr.

The mixture of petroleum-gas and so-called water-gas, which is sometimes used for illuminating, especially in America, may be examined as just described.

A prominent feature of the analysis of such gas is the comparatively large amount of carbonic oxide; and furthermore, high members of the marsh-gas series may be looked for among the illuminants in such gas.

Illuminating gas ought to be supplied to the public in a condition of purity. In its crude state it is loaded with *sulphuretted hydrogen*, and it is useful to be able to measure the amount of this impurity. For this purpose we have designed a special kind of gas-bottle,* which is now well known in certain gas-works, where its practical utility is acknowledged. The gas-bottle is a bottle provided with a hollow stopper designed to contain the reagent (*vide* fig. 3). Its capacity is exactly one-tenth of a cubic foot, or 100 fluid ounces. The hollow

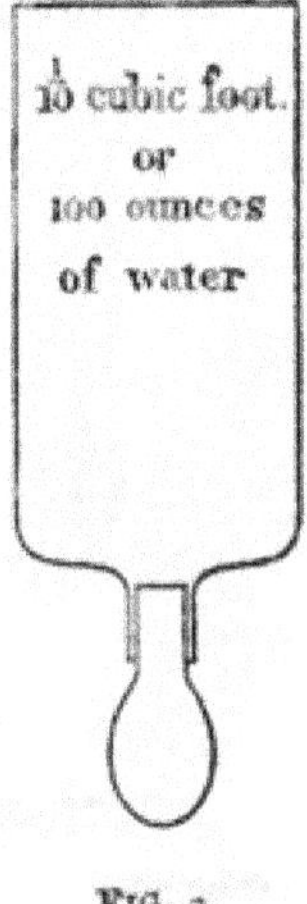

FIG. 3.

* *Philosophical Magazine*, 1881.

stopper is a flask, $2\frac{1}{2}$ fluid ounces, with a neck ground so as to fit accurately into the neck of the gas-bottle.

In order to use the gas-bottle, gas is filled in by *displacement;* that is to say, a tube (*vide* fig. 4) is led up inside to the top of the bottle, and through the tube gas is allowed to flow until the bottle is full of gas. If the gas be issuing at the usual rate (*i.e.,* five cubic feet per hour), a flow of two minutes will suffice to fill the bottle. The use of the flask-stopper will be obvious. It is charged with the reagent, and then inserted into the neck of the gas-bottle. Contact between the reagent and the gas is ensured by giving the bottle twenty vigorous shakes.

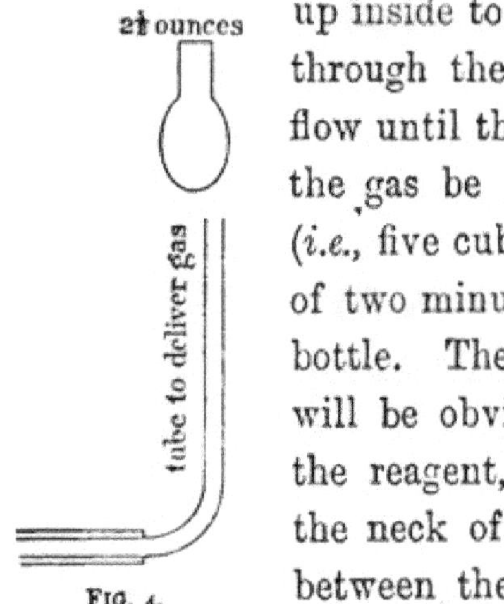

FIG. 4.

Measurement of Sulphuretted Hydrogen (H_2S) *in Coal-Gas.*—For this purpose we require, in addition to the gas-bottle with its flask-stopper, a burette graduated into divisions of $\frac{1}{16}$th fluid ounce; also standard lead solution; also lead-paper.

The burette requires a proper holder (*vide* fig. 5).

The standard lead solution is made by dissolving 35.7 grains of crystallised acetate of lead in one pint of distilled water. Each $\frac{1}{16}$th ounce of the solution precipitates

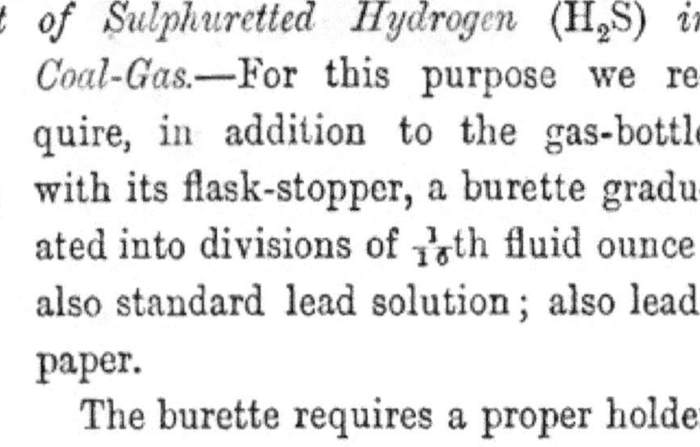
FIG. 5.

$\frac{1}{100}$th grain of sulphuretted hydrogen; and employed with the gas-bottle full of gas, each $\frac{1}{16}$th oz., or each division of standard solution, corresponds to 10 grains of sulphuretted hydrogen per 100 cubic feet of the gas.

The lead-paper is the ordinary lead-paper so well known in gasworks. It is used as an indicator, the cessation of the blackening being the indication that sufficient lead solution has been applied to the $\frac{1}{10}$th foot of gas.

The manner of testing may be illustrated by an example. Suppose there were 275 grains of sulphuretted hydrogen in 100 cubic feet of the gas; the gas-bottle is filled with gas by displacement, twenty divisions of lead-solution are dropped into the flask-stopper, which is then inserted into the neck of the gas-bottle. Twenty shakes are given to the bottle, the liquid is allowed to drain down the sides of the bottle into the flask-stopper, which is then taken out and a strip of lead-paper pushed up into the bottle; the lead-paper is deeply blackened. Five divisions of lead solution are dropped into the flask-stopper, and shaken up with the gas. A trial of lead-paper shows darkening of the paper. Two more divisions of solution are then shaken up with the gas, darkening, but only faint darkening, of the paper. One division of lead solution is added, and the bottle shaken, and then lead-paper refuses to darken. Thus 27 divisions of lead solution is too little, and 28 divisions too much; 27.5 divisions is therefore correct, and 275 grains of sulphuretted hydrogen is the quantity in 100 cubic feet of the gas.

If too much lead solution should happen to be employed in the testing, the operation must be begun afresh.

When there is very little sulphuretted hydrogen in the gas, water must be added to the contents of the flask-stopper, so as to provide sufficient volume to admit of proper contact when the liquid is shaken up with the gas. In practice, these measurements of sulphuretted hydrogen have been found to be easy, certain, and precise.

These examples will serve to illustrate the manner in which testings for H_2S may be turned to account to exhibit the work done by different purifiers:—

Date.	Sulphuretted Hydrogen (Grains per 100 Cubic Feet of Gas).						
	Crude Gas.	Gas after 1st Purifier.	Gas after 2nd Purifier.	Gas after 3rd Purifier.	Action of 1st Purifier.	Action of 2nd Purifier.	Action of 3rd Purifier.
28th Dec.	450	280	100	0	170	180	100
29th Dec.	400	320	160	12	80	160	148

Sulphuretted hydrogen is justly termed an impurity in coal-gas; and coal-gas, in its finished state, when it is supplied to the public, ought to be exquisitely free from sulphuretted hydrogen. That 100 cubic feet of coal-gas (24,000 grains of gas) should not contain 0.5 grain of H_2S is easily attainable in practice.

Measurement of Carbonic Acid (CO_2) in Coal-Gas.

Unlike sulphuretted hydrogen, carbonic acid is not entitled to rank as an impurity or a contamination of coal-gas. It is a diluent, useless for heating or illumination; otherwise it is not harmful in coal-gas. Present in large proportion in coal-gas, it is what by analogy the agriculturist might term a "profligate constituent"—that is to say, it absorbs heat and destroys illuminating power. But in minute proportion in gas there is some doubt whether it is actively destructive of power.

Ammonia (NH_3) in Gas.

As it issues from the retort the gas is loaded with ammonia, part of which passes into the water in the hydraulic main, forming the gas-liquor. The remainder of the ammonia (except the minutest trace) is absorbed in the washers and scrubbers, through which the gas is made to pass. The extraction of ammonia from gas should be as perfect as possible—not alone on account of the value of the ammonia, but because gas contaminated with ammonia acts injuriously upon gas-fittings, and because the products of the combustion of ammonia are not desirable additions to the atmosphere of a dwelling-room.

The detection and measurement of ammonia in gas is accomplished as follows:—

A strip of good yellow turmeric paper, held in a stream of gas containing ammonia, is turned brown by the ammonia. Litmus paper is changed from red to blue by ammonia. Both are in constant use in gas-works for this purpose. In using these test-papers care must be taken to have a sensitive paper, and I should advise the operator to demonstrate the sensitiveness of the paper experimentally before trusting to its indications. The actual measurement of the quantity of ammonia in gas is effected by ascertaining how much sulphuric acid a given volume of gas will saturate. This is a common operation in gasworks.

When only very minute quantities of ammonia are present in gas, "Nesslerising" may be resorted to.

The Nessler test is an alkaline solution of iodide of mercury in iodide of potassium. When the Nessler test is added to a very weak solution of ammonia, it strikes a brownish colour, and the depth of the colour is proportionate to the strength of the ammoniacal solution. Upon this fact is based a very delicate, accurate, and trustworthy method of measuring ammonia.

Bisulphuret of Carbon (CS_2), and in General Sulphur other than Sulphuretted Hydrogen, in Gas.

The presence of CS_2, and organic compounds containing sulphur, is ascertained by burning the gas (which must be free from sulphuretted hydrogen, as shown by its not darkening lead-paper), and obtaining sulphuric acid from the products of the combustion. From the sulphuric acid we may prepare sulphate of baryta, and from the weight of the sulphate of baryta the weight of the sulphur in the gas may be calculated. This is the principle of the method prescribed by the Gas Referees for the measurement of these compounds in coal-gas.

Ten cubic feet of gas are burnt at the rate of half a cubic foot per hour, and the resulting water, holding in solution the sulphuric acid, is carefully collected. The Referees' apparatus in which this operation is performed is here shown (fig. 6).

The gas is measured by means of a small meter before it reaches the jet where it is burnt. The jet, as will be perceived from the figure, is placed within the trumpet-shaped tube whose upper and smaller end opens into the condenser, which is occupied by a number of solid glass balls. When the apparatus is at work, water resulting from the combustion of the hydrogen of the gas condenses on the surface of the glass balls, and runs down the narrow tube into the beaker below. The air required to support the combustion of the gas enters through perforations in the metallic stand on which the trumpet-shaped tube rests. The tube which carries off the gaseous products of the combustion of the gas is prolonged farther than is indicated by the figure—about a yard.

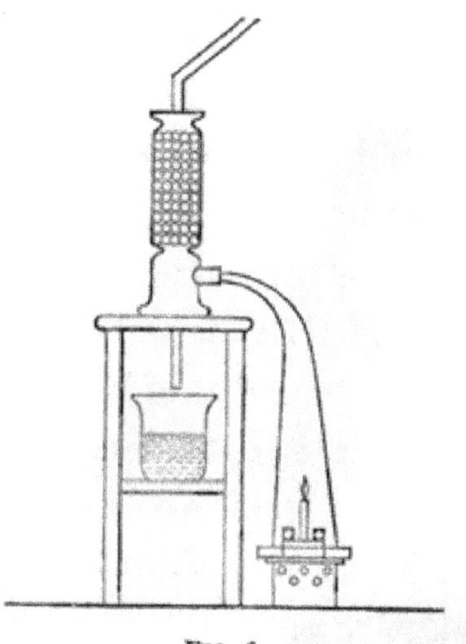

FIG. 6.

After the 10 cubic feet of gas have been burnt, the liquid in the beaker (which contains the sulphuric acid) is rendered slightly acid by means of hydrochloric acid, and then mixed with excess of solution of chloride of barium, which causes a precipitate of sulphate of baryta. The solution is boiled and filtered. The sulphate of baryta in the filter-paper is washed, dried, and ignited, and weighed in a platinum crucible.

The weight of the sulphate of baryta is multiplied by 16 and divided by 116.5 in order to find the weight of the sulphur in the 10 cubic feet of gas.

The Referees recommend that carbonate of ammonia should be placed near the flame, so as to have ammonia in company with the products of combustion. We do not use carbonate of ammonia, but we use iodine instead. We operate on 5 cubic feet of gas, and employ about 20 grains of iodine, which is put among the glass balls in the condenser. The rate at which we burn the gas is at least double that recommended by the Referees.

There is a great objection to burning the gas too slowly, because under such conditions the products of combustion are largely diluted with air, and sulphurous acid is liable to escape condensation. The best results are got by burning at a rapid rate.

The *modus operandi* of the iodine is expressed by this equation :—

$$I_2 + H_2O + SO_2 = 2HI + SO_3.$$

Coal-gas contains from 10 to 50 grains of sulphur per 100 cubic feet. It ought not to contain one grain.

Table of the Weights of Gases.

Weight of 100 Cubic Feet of the following Gases, at 60° F. and 29.9 Inches Barometric Pressure (Dry).

	Lbs.	Grains.
Atmospheric air	7.648	53,536
Nitrogen, N_2	7.430	52,010
Oxygen, O_2	8.491	59,437
Hydrogen, H_2	0.5307	3,715
Marsh-gas, CH_4	4.246	29,722
Ethylene, C_2H_4	7.430	52,010
Carbonic oxide, CO . . .	7.430	52,010
Carbonic acid, CO_2 . . .	11.675	81,725
Sulphuretted hydrogen, H_2S	9.022	63,154

Following Bunsen's example, we have given the weights of the gases as weighed in Berlin, and they will be true for the middle of England, since Berlin and the middle of England are in the same parallel of latitude. Following Bunsen, we have based the calculations on the trustworthy observations of Regnault. We have, however, corrected the results in conformity with Gay-Lussac's law.

THE METRICAL SYSTEM AND ENGLISH WEIGHTS AND MEASURES.

A cubic foot of water measures $6\frac{1}{4}$ gallons, and weighs exactly 1000 ounces. Therefore the same relation exists between the cubic foot and the avoirdupois ounce as between the litre and the gramme.

The usual tables setting out the weight of a litre of a gas in terms of the gramme, may therefore be read as English tables, giving the weight of a cubic foot of a gas in terms of the avoirdupois ounce.

In Bunsen's table we read that a litre of nitrogen gas at 760 millimetres pressure and zero centigrade weighs 1.2566 grammes.

Therefore a cubic foot of nitrogen gas at 760 millimetres pressure and zero centigrade weighs 1.2566 ounces avoirdupois.

Those who are in the habit of making gas-analysis at current pressure and temperature will find it convenient to reduce volumes of gases not to *normal*, but to *average* pressure and temperature, viz., 760 millimetres pressure and 15.6° centigrade.

The following data relating to the common gases will sometimes be found useful.

Table Calculated for Average Pressure and Temperature.

Weight of 1 cubic centimetre of different gases at 760 millimetres pressure (dry) and at 15.6 centigrade:—

	Milligrammes.
Hydrogen, H_2	0.084562
Oxygen, O_2	1.35299
Nitrogen, N_2	1.19387
Carbonic oxide, CO	1.19387
Carbonic acid, CO_2	1.86036
Marsh gas, CH_4	0.67650

SOLUBILITIES OF DIFFERENT GASES IN WATER.

In 100 c. c. of water Bunsen finds that the gases dissolve as follows :—

	TEMPERATURE CENTIGRADE.				
	0°.	5°.	10°.	15°.	20°.
Hydrogen, H_2 . . .	1.930	1.930	1.930	1.930	1.930
Oxygen, O_2	4.114	3.628	3.250	2.989	2.838
Nitrogen, N_2 . . .	2.035	1.794	1.607	1.478	1.403
Carbonic oxide, CO . .	3.287	2.920	2.635	2.432	2.312
Carbonic acid, CO_2 . .	176.67	144.97	118.47	100.2	90.140
Marsh gas, CH_4 . . .	5.449	4.885	4.372	3.909	3.499

INDEX.

PRINTED BY BALLANTYNE, HANSON AND CO.
EDINBURGH AND LONDON.

www.ingramcontent.com/pod-product-compliance
Lightning Source LLC
Chambersburg PA
CBHW020303090426
42735CB00009B/1203

* 9 7 8 3 3 3 7 1 3 9 2 9 2 *